AF270483

DIY SCIENCE FAIR FUN!

CHEMISTRY PROJECT YOUR WAY

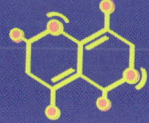

Megan Borgert-Spaniol

Super Sandcastle

An Imprint of Abdo Publishing
abdobooks.com

abdobooks.com

Published by Abdo Publishing, a division of ABDO, PO Box 398166, Minneapolis, Minnesota 55439.
Copyright © 2024 by Abdo Consulting Group, Inc. International copyrights reserved in all countries.
No part of this book may be reproduced in any form without written permission from the publisher.
Super SandCastle™ is a trademark and logo of Abdo Publishing.

Printed in the United States of America, North Mankato, Minnesota

102023
012024

THIS BOOK CONTAINS
RECYCLED MATERIALS

Design: Aruna Rangarajan, Mighty Media, Inc.
Production: Mighty Media, Inc.
Editor: Liz Salzmann
Cover Photographs: Adobe Stock; Mighty Media, Inc.
Interior Photographs: Adobe Stock, pp. 1 (flask), 4 (bottom), 14 (distilled water, seeds); iStockphoto,
pp. 4 (top), 5, 6, 10, 14 (tray), 15, 16 (sprout), 24, 26, 27, 28 (power plant), 31; Mighty Media, Inc., pp. 9
(thermometer), 14 (marker, cups, soil, ruler, spray bottles, tape, vinegar, skewer), 16 (experiment), 17
(experiment), 18 (all), 21 (experiment), 28 (experiment), 29 (experiment); Shutterstock, pp. 7, 8, 9 (both),
11 (all), 13, 14 (measuring cup, measuring spoons), 19, 22, 25, 28 (grass, exhaust), 29 (girl), 30
Design Elements: Shutterstock

Library of Congress Control Number: 2023939283

Publisher's Cataloging-in-Publication Data
Names: Borgert-Spaniol, Megan, author.
Title: Chemistry project your way / by Megan Borgert-Spaniol
Description: Minneapolis, Minnesota : Abdo Publishing, 2024 | Series. DIY science fair fun! | Includes online
resources and index.
Identifiers: ISBN 9781098292041 (lib. bdg.) | ISBN 9781098278946 (ebook)
Subjects: LCSH: Do-it-yourself work--Juvenile literature. | Chemistry--Juvenile literature. | Science projects--
Juvenile literature. | Science fair projects--Juvenile literature.
Classification: DDC 507.8--dc23

Super SandCastle™ books are created by a team of professional educators, reading specialists, and content developers
around five essential components—phonemic awareness, phonics, vocabulary, text comprehension, and fluency—to assist
young readers as they develop reading skills and strategies and increase their general knowledge. All books are written,
reviewed, and leveled for guided reading, early reading intervention, and Accelerated Reader™ programs for use in shared,
guided, and independent reading and writing activities to support a balanced approach to literacy instruction.

CONTENTS

EXPLORE CHEMISTRY

Do you love to watch leaves change colors in the fall? Do you wonder how bread rises in the oven? You might enjoy chemistry! Chemistry is the study of **matter**. Scientists who study chemistry are called chemists.

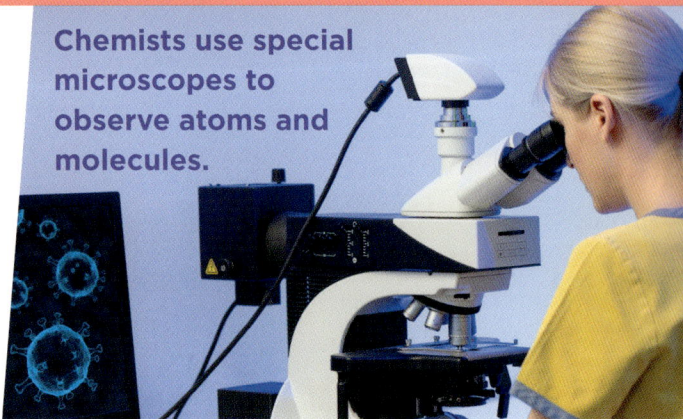
Chemists use special microscopes to observe atoms and molecules.

Solid, liquid, and gas are different states of matter.

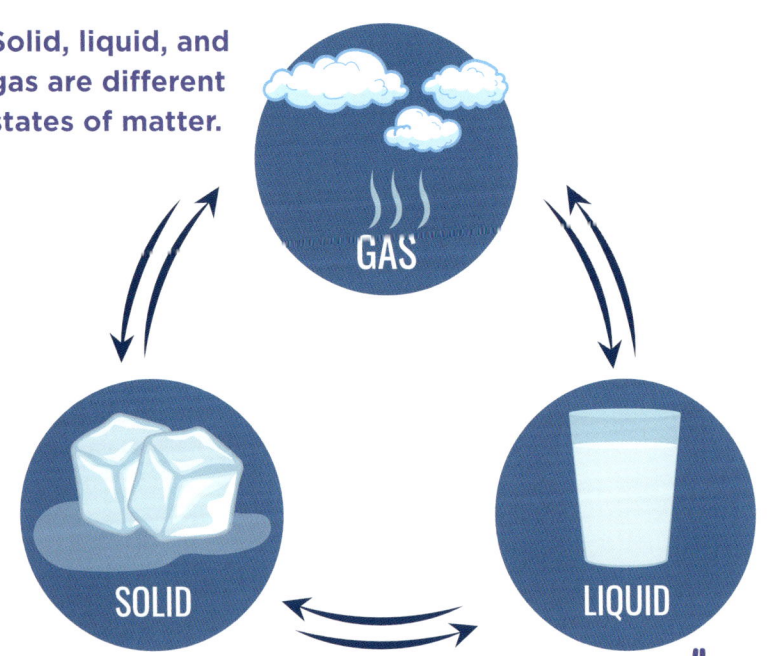

GAS

SOLID

LIQUID

Chemists experiment with solids, liquids, and gases. They study atoms and molecules. They try to find out how matter combines and changes.

H_2O

Chemists use glass beakers, flasks, and test tubes to hold and mix chemicals.

BECOME A SCIENTIST!

Scientists use a process called the scientific method. Check out the steps on the next page. You will use this method to **design** your own chemistry project!

THE SCIENTIFIC METHOD

1 ASK A QUESTION
What would you like to find out?

2 GATHER INFORMATION
What information do you need to understand your topic?

3 FORM A HYPOTHESIS
What do you think is the answer to your question?

4 EXPERIMENT
How can you test your hypothesis to find out if it is correct?

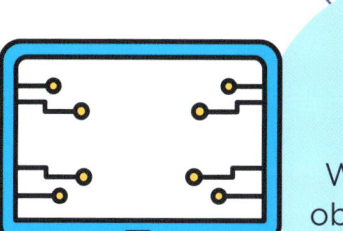

5 RECORD THE RESULTS
What did you observe in your experiment?

6 WRITE A CONCLUSION
Did your results support your hypothesis?

ASK A QUESTION

What topic do you want to learn about?

Maybe you are interested in why rust forms on your bike. Or maybe you want to know more about acid rain. Start asking questions! Write your questions in a notebook so you don't forget them.

What causes rust to form?

What materials are most likely to rust?

GATHER INFORMATION

Maybe you have a lot of questions. That's great!

Scientists often have many questions they'd like to answer. But for now, choose one to **focus** on. Save the others for **future** projects.

It's time to **research** your **topic**. You can gather **information** from many different sources.

▶ **Read online articles about the topic.**

▶ **Read books about the topic.**

▶ **Talk to scientists or other experts.**

What did you learn about your **topic**? Write it down in your notebook. Then you'll have all the **information** you need in one place.

Acid rain is mainly caused by power plants and vehicles releasing gases into the air.

Acid rain is rain that has high levels of sulfuric and nitric acids.

Acid rain harms soil by pulling healthy elements from it.

FORM A HYPOTHESIS

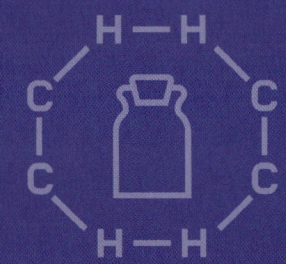

After you research your topic, it's time to form a hypothesis.

Your hypothesis is what you believe is the answer to your question. First, revisit your question. Do you want to change it based on what you learned? Then think of a few different hypotheses. Record them all in your notebook.

QUESTION: How does acid rain affect plants?

HYPOTHESIS 1

Higher levels of acid in rainwater will harm plant growth.

HYPOTHESIS 2

Higher levels of acid in rainwater will benefit plant growth.

HYPOTHESIS 3

Plant growth will not be affected by the level of acid in rainwater.

PREPARE YOUR LAB

Get ready to test your hypothesis.

Find an area with a sturdy table or counter to work on. Then gather the supplies you'll need for your science experiment.

SUPPLIES

distilled water

marker

measuring cup

measuring spoons

3 plastic cups

potting soil

radish seeds

ruler

3 spray bottles

tape

tray

white vinegar

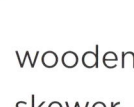
wooden skewer

LAB RULES

All labs have rules that scientists have to follow. Here are some rules for your lab. They will help you stay safe and have fun while doing your experiment!

→ **Ask an adult** for permission to use the materials and do the experiment.

→ **Ask for help** with sharp or hot tools.

→ **Wear goggles** and gloves to protect your eyes and hands.

→ **Clean up** when you are done and put everything away.

STEP 4 EXPERIMENT!

You've gathered the supplies. You've prepared the lab. It's time to experiment!

1 Pour 1 cup (240 mL) distilled water into a spray bottle. Label the bottle D. Pour 1 tablespoon (15 mL) vinegar and ¾ cup (177 mL) plus 3 tablespoons (44 mL) distilled water into another bottle. Label the bottle A for acid. Pour ¼ cup (59 mL) vinegar and ¾ cup (177 mL) distilled water into the third bottle. It has more acid, so label the bottle A+.

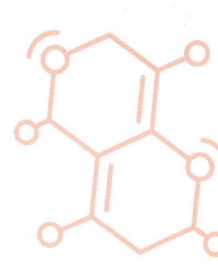

2 Label the cups D, A, and A+ to match the bottles. Fill each cup most of the way with potting soil.

3 Use the skewer to make five wells in the soil in each cup. The wells should be about ½ inch (1.3 cm) deep.

4 Place one radish seed in each well. Gently cover the seeds with soil. Then place the cups on the tray. Set the tray in a warm, sunny spot.

5 Water each cup using the spray bottle with the matching label. Water them at the same time of day. Give each cup 12 sprays each time. Record the number of sprouts in each cup after one week.

LAB TIP

Keep the bottles together in the same place. Then they will all stay the same temperature.

Look at the results of your experiment so far. You might be ready to draw a conclusion. But first, consider any other **variables** that might affect your results.

You planted all the seeds in plastic cups. Then you kept all the cups in the same spot. They were all exposed to the same temperature, **humidity**, and amount of light.

This means the cup material was not a variable in this experiment. Neither were the room conditions.

I watered all the seeds at the same time with the same amount of water.

This means the only **variable** was the amount of acid in the water.

19

STEP 5 — RECORD THE RESULTS

During experiments, scientists record data and other observations. You wrote down how many sprouts you saw in each cup. Now, it's time to record your data to share with others.

Scientists often use tables and graphs. This helps make the results easy for others to understand.

A table organizes **information** in rows and columns.

WATER	ACIDITY	NUMBER OF SPROUTS AFTER ONE WEEK
Distilled (D)	1	5
Acid (A)	2	3
More Acid (A+)	3	0

A bar graph compares data using rectangular bars.

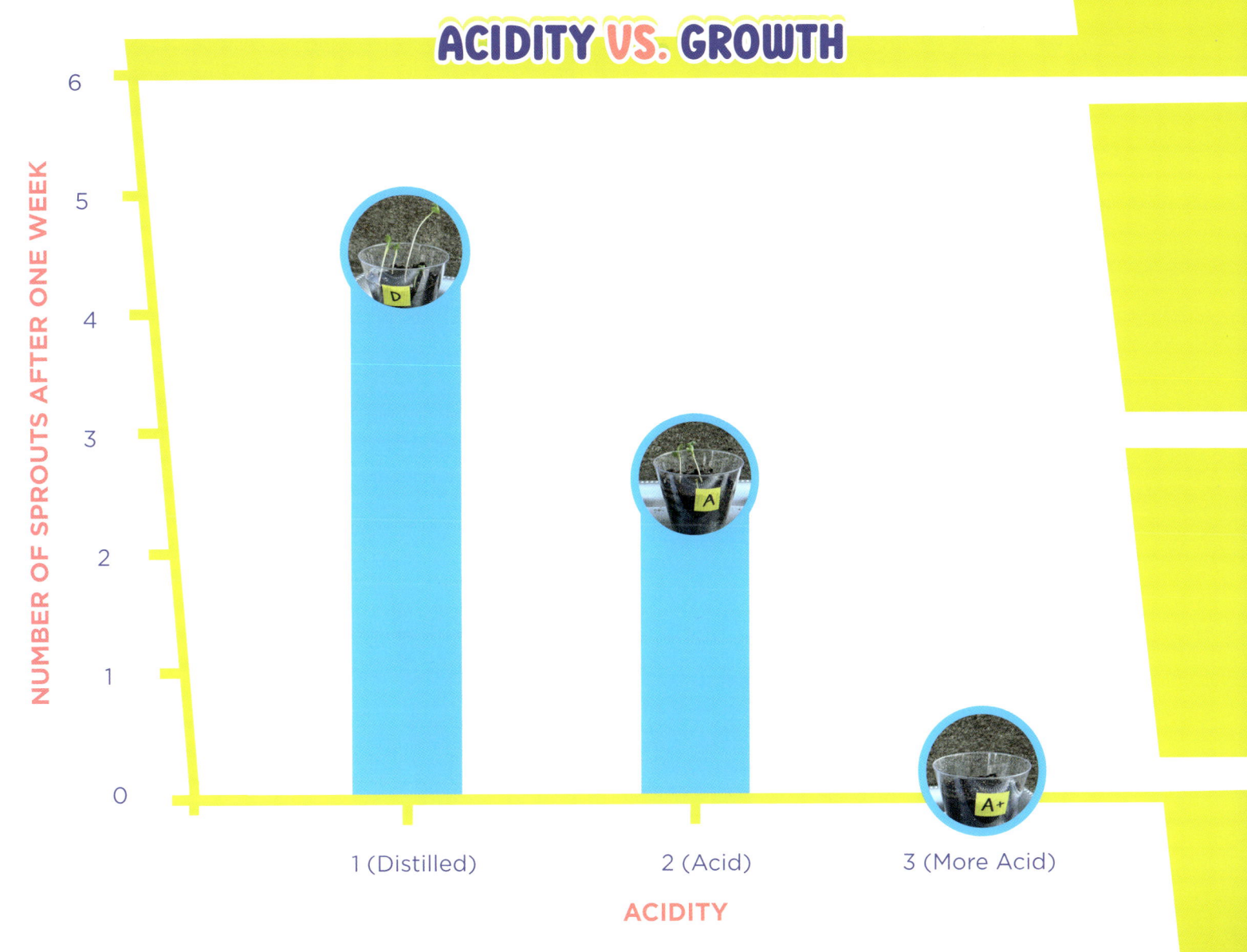

ACIDITY VS. GROWTH

NUMBER OF SPROUTS AFTER ONE WEEK

6

5

4

3

2

1

0

1 (Distilled)   2 (Acid)   3 (More Acid)

ACIDITY

WRITE A CONCLUSION

You recorded your results. Now it's time to write your conclusion. This is a **summary** of your experiment. Your conclusion provides the answer to your original question. It also states whether your results support your hypothesis.

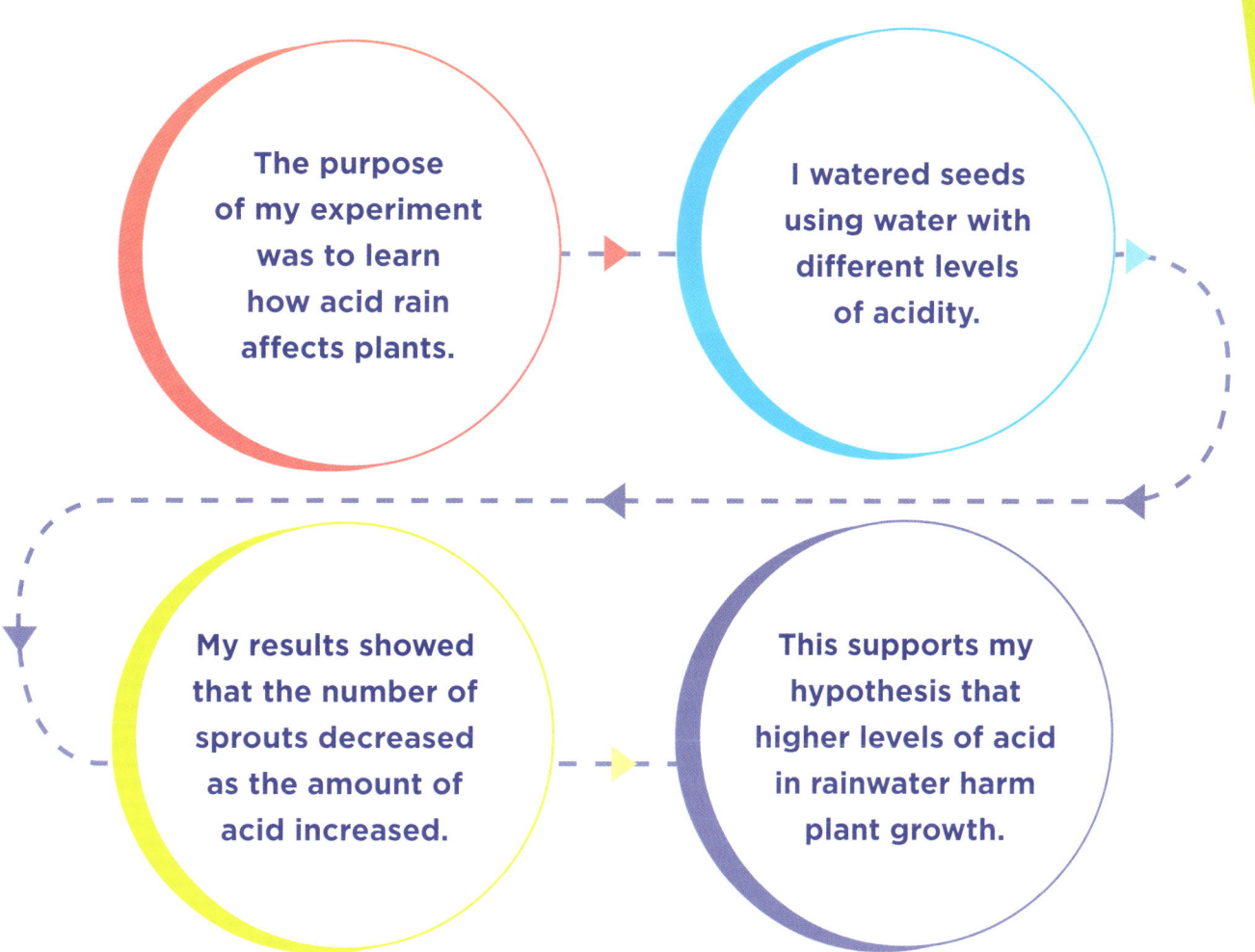

The purpose of my experiment was to learn how acid rain affects plants.

I watered seeds using water with different levels of acidity.

My results showed that the number of sprouts decreased as the amount of acid increased.

This supports my hypothesis that higher levels of acid in rainwater harm plant growth.

Many scientists find that their hypotheses were wrong. There is nothing wrong with being wrong! It means you did your experiment without **bias** and were surprised by the results. That is another mark of a great scientist!

FURTHER RESEARCH

A conclusion offers an answer to your original question. But it can bring up new questions too! For scientists, the end of one experiment often leads to more **research** and new hypotheses to be tested.

What new questions do you have? What could you research further? Do you have a new hypothesis to test?

How does acid rain affect fully grown plants?

How does acid rain affect nonliving things, such as stones and metals?

Does acid rain have the same effect on all types of soil?

25

PRESENT YOUR PROJECT

Young scientists share their **research** at science fairs. Students share what they learned with classmates, teachers, parents, and sometimes judges. It is a chance to show all the work they put into their experiments.

THERE ARE LOTS OF FUN WAYS TO SHARE YOUR NEW KNOWLEDGE!

Demonstrate or show a video of your experiment.

Create comics or other drawings to show your project in a fun way.

Include props, models, or dioramas.

One way to present a project is with a display board. It should show how you followed the scientific method. Turn the page to see a display board of the project in this book!

QUESTION

How does acid rain affect plants?

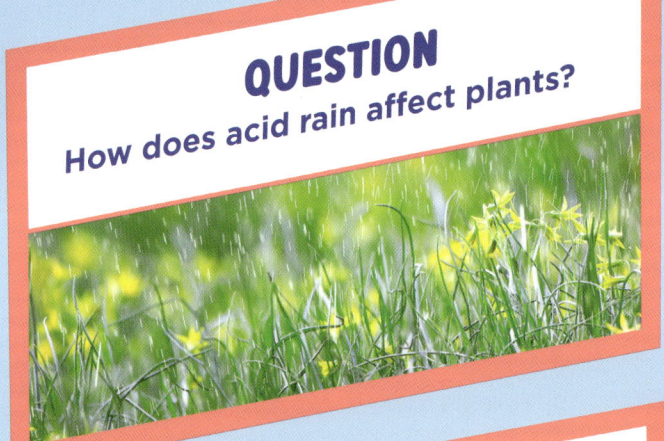

RESEARCH

Acid rain is rain that has high levels of sulfuric and nitric acids. It is mainly caused by power plants and vehicles releasing gases into the air. Acid rain harms soil by pulling healthy elements from it.

Car Exhaust

Power Plant

EFFECTS OF ACID RAIN

HYPOTHESIS

I think higher levels of acid will harm plant growth.

EXPERIMENT

The purpose of my experiment was to learn how acid rain affects plants. In my experiment, I watered seeds using water with different levels of acidity.

No Acid

Some Acid

More Acid

RESULTS

My results showed that the number of sprouts decreased as the amount of acid increased.

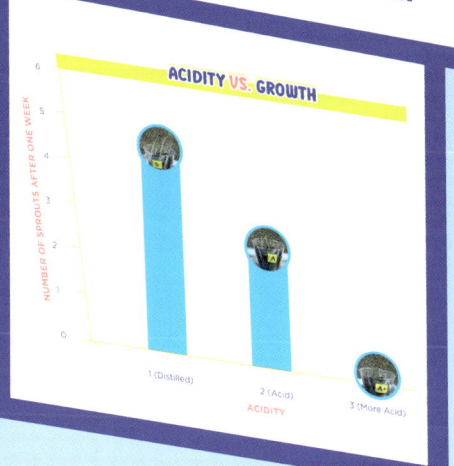

ACIDITY VS. GROWTH
NUMBER OF SPROUTS AFTER ONE WEEK
0
1
2
3
4
5
6
1 (Distilled)
2 (Acid)
3 (More Acid)
ACIDITY

CONCLUSION

The results support my hypothesis that higher levels of acid in rainwater harm plant growth.

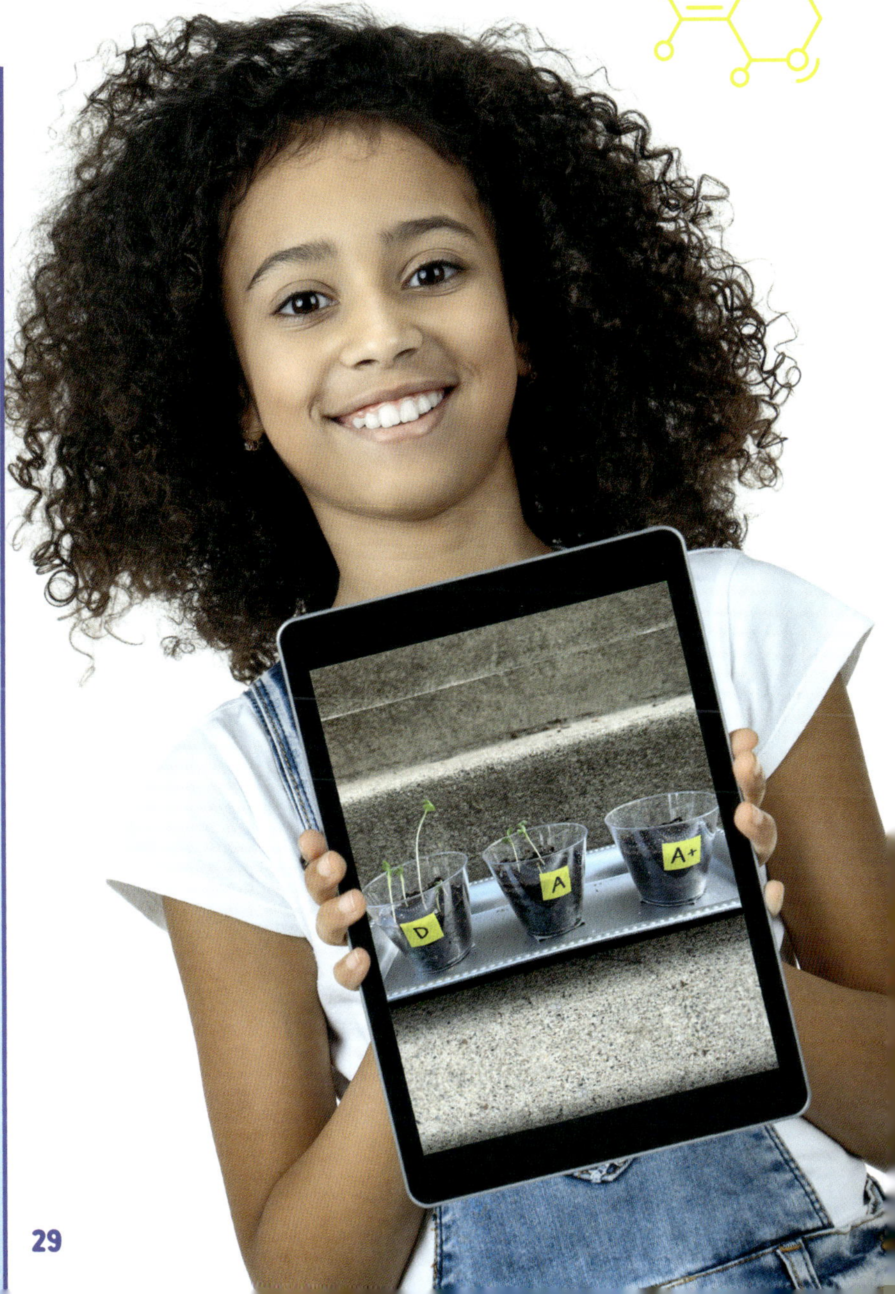
D
A
A+

KEEP ASKING QUESTIONS

Your science project is over. You packed away your display. But don't stop asking questions! What might you do differently if you did the project again? What additional **research** could you do? Is there a related **topic** you would like to explore?

BEYOND THE SCIENCE FAIR

Be a scientist beyond the science fair! You can use parts of the scientific method to find answers to everyday questions. Maybe you have a hypothesis for why your campfire burns out so quickly. Maybe you experiment to learn which type of soap creates more suds. One day, you might use science to do big things. Maybe you'll invent new medicines! Turn your world into a science fair. What will you discover?

GLOSSARY

bias—showing a preference for one result over another.

design—to plan how something will appear or work.

focus—to concentrate on or pay particular attention to.

future—the time that hasn't happened yet.

humidity—the amount of moisture in the air.

information—the facts known about an event or subject.

matter—anything that has weight and takes up space.

research—to find out more about something. Also, a study of something to learn new information.

summary—a short statement of the main points.

topic—the main idea or subject.

variable—a factor in a scientific experiment that may change.